Henkeeping

Jane Eastoe

Henkeeping

Inspiration and practical
advice for beginners

First published in the United Kingdom in 2007 by Collins & Brown.

This edition first published in the United Kingdom in 2017 by
National Trust Books
1 Gower Street
London WC1E 6HD

An imprint of Pavilion Books Company Ltd

ISBN 978-1-90988-199-0

A CIP catalogue for this book is available from the British Library.

10 9 8 7 6 5 4 3 2 1

Reproduction by Spectrum Colour Ltd, UK
Printed and bound by 1010 Printing International Ltd, China

This book can be ordered direct from the publisher at
www.pavilionbooks.com, or try your local bookshop.
Also available at National Trust shops and
www.nationaltrustbooks.co.uk

CONTENTS

WHY KEEP CHICKENS?

Putting good, healthy food on the table is important to most of us but issues such as the closure of local shops, the ever-increasing power of the supermarkets, vast quantities of packaging and carbon footprints make the task a difficult one.

MOST OF US WOULD LIKE to live the good life but, in reality, don't have the time, space or money to cope with the demands it brings. However, it is perfectly possible for us all to make small changes to our way of life. The simple act of keeping a few chickens and keeping your family in eggs is one of the easiest ways to begin. Full self-sufficiency it isn't, but it is a start, a personal contribution to a better way of life. Best of all, you can do it whether you live in the town or the country, even if you have only a small garden.

Chickens are gloriously easy to keep, far simpler and less demanding than most domestic pets; they supply you with eggs and fertilise your garden, which, if you grow fruit or vegetables, is very good news. They regard common garden pests as great delicacies. Moreover, they are delightful company, they follow you around as you garden, scratching through leaves you have raked up and clearing the flower beds of old leaves and debris as they look for food in the soil. They rush to greet you when you return from work and look out for you emerging from the house with treats. Each one has a distinctive personality. Once you have kept hens it is hard to imagine a life without them.

Children adore chickens and will happily take on a lot of the day-to-day care. Furthermore, it will start to teach them where food comes from and the rewards that even small-scale self-sufficiency can bring. Chickens bring life and colour to the garden and everyone will enjoy their antics.

Collecting eggs is a pleasure that doesn't pall with time. Fresh from the nesting box they are surprisingly warm and will vary from day to day in size and perfection. The colour of egg yolks ranges beyond yellow into a warm orange; everything you cook with them tastes and looks that much better. The occasional double yolker is always a pleasant surprise – easy to recognise because it is so freakishly large that it won't fit into the average egg box. Eggs are both delicious and nutritious, containing proteins, carbohydrates and fats, as well as essential minerals and vitamins.

Once upon a time, virtually everyone had a few chickens scratching around in the back garden to supply them with both eggs and meat. As the industrial revolution took hold and the move towards the cities began, the practice of keeping chickens declined. Local shops dealt with the increased demand for eggs and egg farmers looked to improve production. The development of hybrid breeds, which maximised egg production, made poultry farming more profitable and eventually led to the commercial battery farming systems used today.

Scratching around for food in the flowerbed

The blip in this relentless march towards intensive farming methods came during the WWII when food rationing and shortages led the government to encourage people to 'dig for victory' and grow their own. Keeping a few chickens was seen as the only way to avoid the evils of the powdered egg. However when food rationing ceased and eggs were plentiful again the incentive to keep chickens was lost. People wanted to lead modern, convenient, progressive lives and to move away from the privations of their wartime existence. They wanted to buy eggs cheaply in the shops and it was considered to be a little eccentric, if not downright socially infra-dig, to keep chickens.

Now the trend is beginning to reverse. Owning a few chickens is an easy way to take on responsibility for producing some of your own food. Initially it won't save money; not only do you have to purchase your chickens but you also have to supply them with food, housing and, unless you are very lucky, provide a run. However they will repay you handsomely over the years. It is possible to make chicken houses and runs out of all sorts of odds and ends; a lot depends on how fastidious you are about of the look of your garden and the limitations of your budget. If funds will allow there are many glorious chicken houses on the market, with designs to suit all tastes from the contemporary to the rustic.

You need very little space to keep a couple of chickens and even the smallest backyard can accommodate them. If you have a little more room you may be able to keep four or five birds, to ensure that supply meets demand, although two birds keep my family of five in eggs for much of the year – a lot depends on the laying capacity of the breed you select. If space permits, you may even consider buying a dual-purpose breed for both eggs and the table. Chickens do not wreak havoc in the garden, though if you are sensitive about your borders you will need to pick your breed with care.

Why trail to the supermarket, or drive to the farmers' market to buy eggs when you can pick them up fresh from the garden with such ease? Keeping chickens is a fun way to take responsibility for producing your own food; it gives you a taste of what can be achieved and, before you know it, you will be eating very differently.

I confess that we haven't yet grown our own birds for the table, but it is something the family would like to tackle in the future – space permitting. The birds taste very much better than anything you can buy in the supermarket and you know that, although they haven't had a long life (birds are culled for table at six months old), they have surely had a happy one.

A YEAR IN THE LIFE OF A CHICKEN RUN

Whilst it is important to understand the daily routine that chicken keeping entails, there are also tasks that fall outside the remit of daily care. New chickens must be settled in and all birds respond to the changing seasons, so you must be confident that you are meeting their varying requirements.

WINNING APPROVAL

Before you take the plunge and purchase chickens, you would be well advised to speak to your neighbours. They may need reassurance that they won't be woken at the crack of dawn by a cock crowing. If you are just keeping chickens for eggs alone, you do not need to have a cock: hens will lay whether or not their eggs have been fertilised. Hens do not make much noise, their quiet clucking is unlikely to upset anyone and they only screech if they are frightened. You may win your neighbours' approval if you offer to hand over the occasional box of eggs.

You should also examine the deeds of your house and check with your local authority that there are no prohibitions to your keeping a few birds. Generally there aren't; by-laws are designed to prevent people operating a full-scale chicken farm from their back garden, not to prevent you keeping a couple of hens. If you are planning on keeping more than 50 chickens you are legally required to register your flock with DEFRA (details in Useful Organisations at the end of book); however DEFRA is now keen for everyone to register with them, even if they are only keeping two or three chickens. You are not allowed to sell your eggs unless your flock is registered with the local authority and regularly tested for salmonella, however most councils have no objection to surplus eggs being sold over the garden gate.

NEW CHICKENS

The first consideration will be how you are going to transport your chickens. They will need to travel in a box and, whilst you can purchase purpose-built products, a cat box or cardboard box will suffice. Whatever method you choose make sure you put some newspaper in the bottom to catch the droppings. Cardboard boxes must be able to close securely over the chicken's head – you do not want the bird escaping whilst you are hurtling down the motorway. It is very important to cut a number of good sized ventilation holes (at least four slits, three centimetres deep, by ten centimetres long), in the top sides of cardboard boxes – a chicken can easily overheat in such a confined space.

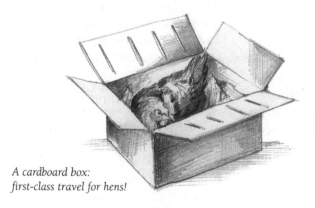

A cardboard box:
first-class travel for hens!

Your chickens will be very nervous when you first bring them home, so it is important that you keep them confined in their run for two or three days until they become familiar with their home base. You may find on the first night that they are reluctant to go into their chicken house and may need a little gentle encouragement. Once inside, they may then be reluctant to emerge the next day. Give them time to build up confidence, but if the day proceeds and they have still not emerged for food and water you may have to give them a gentle shove to get them out of doors. You should find that they make their own way back into the chicken house on their second night. If you are going to let them roam free in the garden, limit their time out initially and allow them to get used to the space, it helps build up their confidence – and yours – if you stay out in the garden with them. They will very quickly relish their freedom and pace the run in fury when you don't let them out.

If you are introducing a new chicken to a flock be prepared that the newcomer will not be welcomed with open wings. Chickens operate a strict pecking order and whenever a new bird is introduced – even if you are keeping only two birds – the established chicken will want to either maintain pole position or to step into it. The general advice is to wait until dark and pop the newcomer into the chicken house, when the other birds are settled. In practice, I have never found this works – the newcomer has clearly suffered overnight and spends its time being mercilessly chivvied in and out of the house by the older birds the next day.

Hens can very quickly hurt a newcomer and will peck repeatedly at her rear end; if they draw blood the bullying will only intensify as chickens are drawn to the colour. My hens roam free, so the best method for me is to put the new bird in the chicken house and remove the older hens – take them out when it is dark, pop them in a cat box and put them somewhere dark and cool in my house overnight (they will also enjoy a calming stroke in the darkness). Try putting them all together at night after two or three days.

In the daytime, the newcomer should initially be kept in the run while the established birds range free outside the run (the upset usually stops them laying for a few days anyway). This allows them time to get accustomed to the interloper without being able to torment her or keep her away from food or nesting boxes. Repeat this process for a few days before allowing them to be together; start by releasing the newcomer into the garden with the older bird. The established hen will chase the newcomer and this may go on for a few weeks, but little by little you will see them getting closer and closer, and soon they will be inseparable friends. One will be established as Top Chicken and will make all the heavy decisions – where to dust bathe, when to hide in the bushes at the top of the garden and when to sunbathe on the front door step.

PETS

Chickens function pretty well with most domestic animals. Cats are generally no problem at all; a few inquisitive sniffs and the chickens will quickly let the cat know who is boss – the chickens! My two cats, who are both ferocious hunters, slink past the hens, giving them a very wide berth. Occasionally, if they pass too close, the chickens will chase them just to remind the cat who is the more powerful. However some cats are chicken killers: it's best to keep a

The hen can rule the roost

watchful eye open when you first get your birds.

Many dogs are perfectly at ease with hens. We have had all sorts of breeds and thus far have had no problems. However, some dogs will not tolerate the chickens and will kill them. If you know anyone who has chickens take your dog to see them – keep the dog on a lead – and watch how your dog reacts. (If you keep chickens and a dog comes to visit, ask its owners to keep it on a lead.)

Introduce your pets gradually and keep chickens enclosed in the run until you feel confident enough to let them loose with your pet. Keep a close eye on all parties initially and limit the time they are in the same space.

Chicken proofing/garden protection

Free-range chickens are the most content birds, they lead
happy and busy lives. However, if you let your chickens out of
the run you must be prepared to ensure they cannot escape
from your garden – however lovely their space they are
curious creatures and long to explore. They will plot
expeditions into the neighbour's garden, or into your house to
explore uncharted territories. If the garden is fenced all
around this is less of a problem, but you can always fill holes
in hedges or gaps in fencing very inexpensively with pieces of
chicken wire. This can easily be wired or stapled into position.
Some breeds fly better than others, the heavy and fancy breeds
tend to be hopeless at staying airborne; fences of just a metre
high will restrict their movements. Remember, if a chicken has
escaped, she will usually be able to return of her own accord –
and will do so as it gets dark. However, your neighbours may
not appreciate visitors and if they have pets of their own, your
birds will be vulnerable.

Wing Clipping

Some breeds are very flighty if kept free range and may need
to have their wings clipped others are less so, but may
surprise you as their confidence increases. My birds are
contained with a one metre high wall as long as I clip their
wings. This is an easy task; it takes just a few minutes and
only has to be done annually. Wing clipping does not hurt the
hen as long as it is done correctly – a bit like us cutting our
nails. You trim the primary flight feathers of one wing only. It
can be helpful to have someone with you to do this process.
They can hold the bird whilst you spread out its wing and,

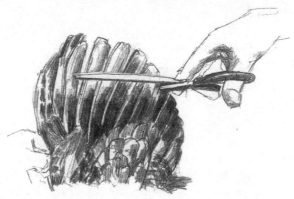

*Clip the primary feathers
on one wing only*

with a sharp pair of scissors, cut the primary flight feathers
(the longest feathers at the tip of the wing, around ten in
total). This prevents the bird from flying. Of course when
chickens have moulted this procedure must be repeated.

COLD WEATHER

Chickens are robust birds and they can cope with most
extremes of temperature – though they hate wind and snow
and look miserable and bedraggled in the rain. If they are kept
in a run, ensure that they have shelter from the wind and rain,
if they roam the garden they will find their own, usually under
a dense and leafy bush. My two girls like to stand on our
sunny, sheltered front doorstep, rain or shine. On bitterly cold
winter days you must check that their water supply is not
frozen. Similarly you may want to rub a little petroleum jelly
onto your hen's comb to protect it from the cold and stop it
getting frostbite. Expect your chickens to provide fewer eggs
over the winter months; the shorter days impact on laying as
light is crucial for egg production.

Broody birds

Birds are called broody when they attempt to hatch their eggs. Of course if you don't have a cockerel your hen can sit on her unfertilised eggs forever and they will never hatch. You will recognise the signs of broodiness, hens refusing to leave the nest, developing bare patches on their chests and becoming more aggressive towards their sisters and to you. Some breeds tend more toward broodiness than others, obviously commercial egg producers will look to breed this trait out of their high-performing hybrids as it creates problems for them. Similarly, certain breeds are used as sitters for others who are not such good mothers. Birds tend to get broody in the spring after their first birthday and as the days lengthen. Collecting eggs promptly is one of the simplest ways to discourage broodiness. However, if you do get a broody hen, put her in a special house with a slatted or wire floor and no nest box, she will find it difficult to squat and this will discourage her inclination to motherhood. It may be two weeks before she resumes laying.

Sitting birds

If you have a broody hen and decide you want to have the full lifecycle experience you have a range of options. Obviously, if you have a cockerel you can let nature take its course, if not you can obtain eggs from breeders that you can slip under your broody hen. Your hen must be kept separate from the rest of the birds as she will be hogging the nest box. Birds usually sit on 9–15 eggs at a time, for 21 days in total, turning them regularly and keeping them warm. During this period you may well have to prise her away from the nest – shut the door to the nest box and make sure she takes food and water and a break for up to half an hour.

When the chicks hatch they will still need to be kept separate from the rest of the flock until they are around eight weeks old. Their mother will teach them how to feed and scratch and will generally take very good care of them. They will need to be fed chick crumbs and you will have to put a handful of stones in their water basins to stop them falling in and drowning.

If you want to start the chicken-keeping process from scratch you will have to play mother and will need an incubator for the eggs and then a brooder (a special box usually heated by an infra-red light) to keep the chicks warm and safe once they are hatched. Heat is reduced by two to three degrees each week until the chicks are six to eight weeks old and have enough feathers to keep them warm.

Here Betty Macdonald gives the low-down on heavy duty chick rearing in 1930s America in her book *The Egg and I*. "It all began with the baby chicks, they came first while I was still very pregnant and getting down on my hands and knees to peer under the brooder at the thermometer was a major undertaking. Bob and I scrubbed the brooder house, walls, floors, even the front porch with Lysol and boiling water. The brooder house had two rooms, the brooder room and the cool room. In the brooder room we had two coal-oil brooders, which we lit and checked temperatures on a week before the chicks arrived. The brooder floor was covered with canvas and peat moss and had drinking fountains and little mash hoppers scattered here and there. At last the chicks arrived and Bob drove to Docktown and returned with ten cartons with air holes in their sides, in each of which yeeped 100 chicks. We stacked the cartons in the cool room and then, one by one, we carried them carefully into the brooder room, took off the lids and gently lifted out the little chicks and tucked them under the brooder, where they immediately set to work to suffocate each other.

"From that day forward my life was one living hell. Up at four – start the kitchen fire – put the coffee on – go out to the baby chicks – come back and slice off some ham and sling it into the frying pan – out to the baby chicks with warm water – put toast into the oven – out to the baby chicks with mash – set the breakfast table – out to the baby chicks with chick food – open a can of fruit – out to the baby chicks again, on and on through the day. I felt as though I was living a nightmare, fleeing down the track in front of an onrushing locomotive. I raced through each day leaving a trail of things undone."

MOULTS

Chickens lose many of their feathers annually and in quite a dramatic way. It is very worrying when you suddenly notice that the chicken house is full of feathers and that your beloved bird is looking somewhat bald – but don't worry, it's perfectly normal. Birds lose their feathers in sequence beginning with the neck and the back, moving on to the breast and the petticoats and finally on to the back and tail. A young and perky bird will take around about six weeks over her moult, but an older hen can take up to three months to refeather.

There is no specific time of year that this happens: it depends on when your bird was hatched. It is particularly distressing to watch when it occurs in January and February, just when birds most need their feathers for warmth. You will first notice that your chicken is bald at the back of her neck and that there are tiny white tubes all over her neck. Next you will be surprised to see that when the wind blows her skirts up she will look as though she has forgotten to put any knickers on. By the time the tail feathers are falling she will look terrible.

Poultry farmers will often withhold a bird's food for 24 hours, or move them on a less nutritious feed as this encourages the feathers to fall out faster; the bird is then moved onto concentrated foods to give it the nutrition it requires to replace feathers quickly. Supplements are readily available and will give your bird a boost. The first sign that feathers are growing back is when you can see tiny clumps emerging from the tubes at the back of her neck.

The real downside to the moult, aside from the feathers all over the garden, is that your hen will probably stop laying. Perversely my chickens all moult at the same time – November – and I am forced to buy free-range eggs from the shop. They do not taste anything like as wonderful as our eggs and it also means we rarely have our own eggs for Christmas, which doesn't impress our visitors at all!

HEALTH CHECKS

It is worth getting into the habit of checking your birds over at regular intervals to see if they are healthy or suffering from any problems. Look at the feathers around the vent and under the wings to see if the chicken is suffering from mites; similarly check that her legs are nice and smooth. Look at the colour of her comb, feel her crop and keep an eye on the condition of the droppings – all can indicate health problems. Also check out the chicken house when you clean it for signs of infestation. For more information on chicken health see the chapter on Problems, page 76.

Washing

Chickens clean themselves, they have a daily preen in the morning to ensure that their feathers are all gleaming, waterproof and in place. However, fancy breeds can get horribly mucky in some conditions and need a helping hand. Similarly if a bird has a severe infestation of lice or mites, a wash may help remove them. You will need to prepare a bowl or bath of water that is warm, not hot, to the touch. Make sure that all the feathers are wet by stroking the bird essentially the wrong way – feathers are designed to be partially waterproof and are coated in a protective film so you do have to work the water in. You can use a very mild baby shampoo to wash the chicken but be very careful that it doesn't drink the water and keep the detergent away from its face. Rinse the bird thoroughly to remove all traces of shampoo. The chicken can then be dried with a towel or given a gentle blow dry. You must ensure that the bird is completely dry before it is returned to the chicken house. It takes a week for your chicken to re-instate the protective film over its feathers so make sure it is kept somewhere dry. If your chicken's legs are very dirty they will appreciate you giving them a clean with a cloth and some warm water – although you will be rarely called upon to perform this task.

PREDATORS

The fox is the main predator of the chicken. Dogs can be a nuisance and ferrets can terrorise your birds. However, rats and mice can also upset them. A well-designed chicken house and run will keep the fox at bay, see chapters on Hen Houses and Runs and Problems for more information. Mice and rats will be discouraged if food is kept in secure bins and therefore inaccessible, similarly if eggs are collected daily. For bio-security reasons, contact with wild birds is to be discouraged – not that this is a realistic proposition if you keep your birds free range. However, you can ensure that wild birds are kept away from your chicken food and water and also keep the chicken's feeder and drinker under cover in the run, so that wild bird droppings cannot contaminate it.

HOLIDAYS

You will be surprised how delighted people will be to look after your chickens overnight or for few weeks whilst you get away from it all. It gives them a tiny taste of self-sufficiency and, the one pay off they should always get, is to collect and keep the eggs. Neighbours are obviously ideal chicken sitters, particularly for overnight trips. You can give your birds extra feed and water and then it is simply a case of someone shutting the door after the birds have retired for the evening and then letting them out in the morning. Longer stays require a little more planning. Often the best solution, assuming you have only a few chickens, is to take your chicken house and run to a friend – it's much more convenient for them to have it all on their doorstep instead of making twice daily trips to you.

BASIC CARE

Chickens are extraordinarily easy to care for – so much so that once you have them it is hard to understand why everyone else isn't keeping them too.

HAVING BEEN THROUGH the standard pet menu with my own children – guinea pigs, rabbits, hamsters and rats – I can emphatically state that chickens are much easier to look after and a whole lot more rewarding too. The eggs your chickens produce will taste better than any eggs you have ever tasted before and there is nothing so satisfying as being able to pop out into the garden to get a few freshly laid eggs.

The basic principles of care are simple. In the morning you let the chickens out of their house and check and replenish their food and water supply. They like to be let out when it is light – so in summer it is a job you do when you wake up – and, yes, they do cope with a later start at the weekend, though I swear they can be heard grumbling! In my house the rule is simple; whoever is first up puts the kettle on and then nips out to the garden to let the chickens out and check their food and water. In winter it is a case of waiting for it to get light enough to let them out. At some point in the day you check for eggs – if you haven't got around to it earlier in the day you can always do this when you shut them up for the night. Try not to leave eggs in the chicken house overnight, the chickens may break them accidentally – or deliberately – as some chickens do develop a taste for eggs. Either way it makes the weekly clean of the chicken house more problematic and your egg supply suffers.

Your chickens will develop a routine of their own for

egg laying – the times of day they lay can vary throughout the year as the laying process is affected by light levels (poultry farmers put artificial lighting in the chicken houses in winter to maintain production levels). At some point in the day, most commonly after they have had breakfast and taken a stroll, the chickens will return to the hen house and lay. Hens like to take their time over this activity; they cogitate on the nest box and will rearrange the straw, all the while singing away to themselves. It is a real joy to listen to your chickens chuntering away and you can usually tell when the egg is about to drop as they get far more vocal. My oldest chicken, Wilma, generally spends a good hour laying her egg.

Hens do not like being disturbed when they are laying. If you ever do so, you will hear the chicken version of an outraged scream, and they are quite likely to take umbrage and climb out of the nest in fury! Chickens really are best left

Free-range birds will nest in the garden

alone to perform their daily miracle in peace and quiet.

If your hens are free range then you will find that they may stop laying in the chicken house, particularly in the warm summer months when there is plenty of cool, leafy vegetation behind which they can conceal themselves. They will continue to lay in their own secret nest until the eggs are removed, when they will find a new nest! Therefore, if your chicken starts to lay in the garden, the great trick is to never remove all her eggs at once. Either leave one there permanently – mark it 'Do Not Eat' and return it to the nest – or leave a rubber or marble egg there instead.

If you don't know where your chicken is laying then try to watch her carefully to see where she goes. If she vanishes, go out into the garden and roam around listening carefully and you may hear her distinctive laying coo. Your other chickens can also help to reveal her hiding place – pairs tend to loiter quite closely while their friend is laying. One of our chickens had, apparently, stopped laying one warm summer and it took me some time to find the location of the nest. When I did so there was massive pile of 13 eggs. It had reached such heights that it was a miracle she was able to clamber on top to lay any more.

Chickens put themselves to bed in the evening; their eyesight is terrible and they cannot see as darkness falls, so they naturally retire to roost at dusk. At this point it is easy to lock them in safely for the night. Once again, in winter this is a task that is done as you return from work – in summer, with the long light evenings, it is all too easy to forget to lock them up.

Hens are at their most vulnerable to predators at night – this is when the fox will pay a visit – so generally they need to be shut up for their own safety. However, depending on the physical security of your chicken house and run arrangement, if the chickens are secure without being locked up in their house then, in summer, they will appreciate the door to the house being left open – they will certainly be out and about in their run much earlier with this arrangement. In winter they benefit from the door being locked to help keep out the draughts.

FEEDING YOUR CHICKENS

Chickens enjoy a varied diet; left to their own devices they will eat seeds, plants, fruit, berries, insects, worms and grubs. The fantastic news for gardeners is that chickens even regard snails as a treat, although admittedly they prefer them small. However, even free-range chickens should have fresh chicken food provided daily as this supplies all the essential vitamins, nutrients, protein and carbohydrate that they require.

All kinds of containers can be used to hold food, although they should have a little weight to them (flimsy bowls will be tipped up in seconds) and should be easy to clean. Commercial feeders release a steady supply of food to the bird, but do not allow them to tip it up; the design also discourages them from flinging too much of it around. If too much food is spilled on the floor of the run it will attract mice and rats, so there are advantages to purpose-designed feeders. These are not generally available from pet shops so you may need to have a look on the internet. Alternatively, take a day to enjoy the delights of a poultry show where you will find all manner of tempting specialist equipment to spend your money on.

Commercial chicken food for laying birds contains oats, barley, wheat and maize. It should also contain protein in the form of soya. In some commercial mixes the protein content is not necessarily from a vegetable source and this is not ideal. If you want your chicken to have a natural diet then check the labels on chicken food before you buy. Organic chicken food is also available.

Chicken food comes as pellets or mash. Pellets are compressed into small pieces, this ensures that the chickens eat everything and can't be selective. Mash can be given dry or mixed with water. Fancy chickens fare better with dry food as they trail their exotic feathers in wet mash and spoil their good looks. Depending on their size, laying hens tend to consume between 30–100 grammes of chicken food daily. Don't worry too much, you will quickly discover whether you are over or under feeding when you check the food bowls in the morning. You can try the different types of food and see which your chickens prefer. As a general rule, free-range chickens do not overeat, they derive enormous pleasure from finding their own tit bits.

Chickens also enjoy mixed grain; corn, oats, wheat and barley and they adore sunflower seeds. These should be given as an extra treat and not as their main diet. My two hens, Wilma and Dora, used to dash out of their run every morning and race to our wild bird feeder; they would position themselves underneath and pass a contented half hour picking up tasty morsels that the wild birds had dropped. However, we have now moved the wild bird feeder as DEFRA advises that contact between poultry and wild birds is minimised.

The 'girls' worship my husband; rushing to him as soon as he sets foot outside the door. He is the one who is responsible for filling up the wild bird feeder and always gives them their own treat – a separate handful of seed. Store your chicken food in a dry place and ensure the container has a lid on it to keep out damp, dirt and pests.

Chickens need plenty of greens, if your birds are penned in a run on grass, you may want to move it every day or so to ensure they have a constant supply of fresh grass – they will work like efficient mowing machines. You can offer a fresh supply of greens daily – lettuce, spinach, cabbage, etc. – in a string bag suspended from the run. It's good for them and keeps them happily occupied for a while; similarly you can get special mangers which will allow the birds to peck away but will keep the greens off the dirty floor.

Chickens scratch at the grass, though moving the run daily minimises this problem. Nothing will save your turf however, when the chickens discover a rich seam of goodies underneath. Crane fly larvae (*Tipula*) for instance; the birds will scratch and scratch to find the wretched things until the grass is destroyed! My chickens are free-range for most of the year and they do not damage the lawn too much, although I have to pen them up if the winter is very cold – the fox gets desperate – and then they do inflict more damage on the garden, but it always recovers in spring.

Chickens also love scraps from the table and will happily eat leftover toast crusts, cereal, salads and some leafy vegetables (different birds have different likes and dislikes). They do get tired of some foods and whilst they consume porridge with relish in October, by January they turn up their beaks at it. Grapes seem to be a favourite delicacy of all birds and they adore an occasional piece of stale cake. You should not give hens animal protein as this is not a food that they would obtain naturally. Don't give them any strongly flavoured leftovers, such as curry for instance, as the taste will find its way into the egg.

Chickens love a treat and they love you for providing it. Wilma and Dora will come when their names are called – actually come is rather sedate a description. Our girls do a lopsided sprint across the garden, clucking loudly with wings flapping, even taking off for a moment or two, both equally desperate to be the first to get to the potential treat. They are never too tired or too jaded to respond to any summons with less than excessive enthusiasm and excitement. On such occasions one understands why the term hen party was coined.

Also essential to egg production is grit. Mixed grit supplies the laying bird with the calcium it needs to produce eggs. Commercial grit is made from ground oyster shells. Granite grit (tiny stones) is also required to help the chicken digest food. Granite grit and mixed grit can be combined in containers kept close to the nest boxes so that hens can peck at it as they lay. My garden is full of tiny stones and, as we live close to the seaside, it is also awash with seashells brought back from the beach by the children. These collapse over the course of a few years and my chickens seem to thrive on obtaining all their grit direct from the garden.

Birds require different foodstuffs at different times. A bird laying to produce eggs for hatching requires food with a higher mixed grain – content than one laying eggs for consumption. A chick will need chick mash or pellets and some chick crumbs, as they can only manage small grain and seed. At around six weeks old growers' pellets should be substituted until the bird has laid its first egg, when it can move on to regular food. Birds may need supplements when they are moulting – see Moults, page 19 for more information.

WATER

Chickens must have a good supply of fresh water. If a laying hen runs out, her egg production can be affected for a few days. Birds, like us, require more water in the hot months, so refresh their water daily. Some supplements are given via the water supply.

There is a lot to recommend purpose-designed drinkers. Chickens can seem alarmingly stupid at times and, when they douse themselves in water on a daily basis by standing on the rim of their bowl, or paddling in it, you really do begin to question their capacity to learn. Commercial drinkers will make sure the birds cannot tip their water – or indeed dive into it! If you plan to house a fancy breed you will need to select the drinker carefully – they will not be pleased if their feathers get soaked every time they take a drink. If you have chicks you will need a proper drinker as these babies have a tendency to drown in the smallest amounts of water.

CLEANING

The main chore in day-to-day chicken care is keeping the chicken house clean. Your chickens will be healthier and happier if they have a clean environment to sleep and lay in. Depending on the design of chicken house you choose, it should be possible to remove excess mess from the droppings tray daily to keep the house smelling sweet. In my case, this a 20-second procedure – slide the tray out, tap the back so that the droppings fall on a nearby flower bed or compost heap then slide the tray back in position.

Once a week the chicken house should be thoroughly cleaned; old straw should be discarded, the roosting bars and the nest box given a light scrub and a hose down until everything smells fresh. Lice tend to live in the crevices at the ends of roosting bars so leave these out to dry in the sunlight as the mites can only survive in darkness. In summer I leave the house open for several hours to dry and air – you can dry it with old newspapers to speed the process in winter. Then I add fresh straw and put the house back together.

Once clean and reassembled, you will find your chickens immediately dive inside their home to check that everything has been cleaned to their satisfaction. It usually takes me about 15–20 minutes to give the chicken house its weekly clean. Do check where your chickens are before you embark on this process – otherwise you might surprise a hen in the middle of laying.

Every month to three months the interior of the chicken house should be given a more in-depth scrub with disinfectant. I leave my roosting bars to soak in dilute disinfectant for a little while to ensure they are really clean.

DUST BATHS

When you allow your chickens to roam it will not be long before you come across them lying on their sides in a dry piece of soil, apparently delightedly showering themselves and each other in dirt. This is called a dust bath and is the method by which chickens remove mites and lice, as well as cooling themselves down. When they finally get up and shake the dust from their feathers some of the pests go with it. Wilma and Dora have a number of favoured spots for dust bathing, all of which allow them to lie side by side – bathing seems to be a social event. They move around their various dust baths, often apparently choosing the spot according to the position

A dust bath – a hen's beauty treatment

of the sun. Sometimes they like to warm themselves as they bathe, but in the height of summer they will choose a shady spot. They even have a dust bath under a bush, which they use when it's raining.

If you don't let your chickens out to roam on a daily basis you should provide them with a dust bath. Indeed, you may want to do this anyway to discourage your chickens from scratching holes in the flower borders. Use a shallow tray filled with sharp or play sand, fine dry soil and ash, or a combination of the three ingredients and the chicken can bathe to her heart's content. If you can provide a dust bath with a little roof, so the contents are always dry, then so much the better.

MESS

Whilst chicken mess is an excellent fertiliser, it should ideally be left to rot down in the compost before it is put onto the flowerbed; it can burn plants if applied directly. However, having said this I must be honest and confess that my daily chore in spring and summer is to work around my paths and lawn, hoe in hand, flicking the mess onto the flower beds like an eccentric form of golf. I am quite skilled at this now and rarely overshoot my target. Using the hose is another easy way of removing droppings (when there isn't a hosepipe ban). Simply aim the jet of water at the offending deposits and they will vanish into the soil. This is an essential task if you don't want to be walking mess into the house and also if your children, like mine, are keen garden football players.

Handling a Chicken

Chickens are not naturally cooperative so it is worth getting to grips with a little chicken psychology to enable you to persuade them to do the things that you want them to. Chickens are flighty and nervous and you can easily frighten them; the more frightened they get, the harder they will be to control and their egg production may ultimately be affected.

Chickens need calm handling. When you want them to go into their run the most foolproof method is spread your arms very wide and move toward them slowly but steadily – not getting too close. Steer them back to their run – it can be helpful to throw something delicious in at the last moment so that they race in for a treat. Tempting hens into their run with food works part of the time, but when they have already made up their minds to do something else they will simply ignore your summons.

Encourage your chickens to get used to you and to see you as a source of treats. Crouch down and encourage them to eat out of your hand. They probably won't do it immediately – give them time to become more confident. When they are happy to eat out of your hand you can try stroking them on

33

their backs, but avoid any sudden or jerky movements. Some chickens simply don't like being stroked and they will always avoid your touch; some breeds are very docile and will enjoy being handled and quickly become tame. On a general basis, the heavier breeds are gentler than the lights and certain breeds of bantam and fancy chickens make delightfully amiable pets.

As hens become accustomed to you, they will crouch down on the floor and brace themselves for action. Stroke their backs and they will be very content to stay put for a little while – this is the perfect time to pick them up. Put your hands either side of the body around the wings, pinning them

Hold wings gently but firmly to stop flapping

to her side as you lift, to stop her flapping wildly. Hold her against your body, her head can face forwards or backwards. Keeping a firm grip on the free wing with one hand, use your other hand to slip your fingers around and between her legs to get a good grip. She is now comfortably supported, she will try to flap but talk to her gently and stroke her.

The chicken crouch routine happens more frequently in spring and summer, but when she refuses to crouch down obligingly, try bringing your hand down on her back from above and press down gently. If she resists this technique get someone else to try to help you – the task is always easier with two. Don't ever pick a chicken up by its legs.

BREEDS

*Chickens come in a bewildering array of shapes and sizes.
There are chickens with crests, beards and muffs; chickens
with pencilled or spangled marking patterns – all available
in exotic colours from lavender to silver.*

CHOOSING A BREED can be confusing but don't be daunted.
It is worth taking time to read about a few of the main
breeds to determine what best fits your particular
requirements. Some breeds are strong layers, others are good
table birds, some are better with children while some breeds
are kinder on the garden than others.

As with all aspects of animal, fruit and vegetable
production, intensive farming has endeavoured to create
hybrid breeds and varieties that will produce a bumper
return. However, it is essential for the health and wellbeing
of all species of plant and animal that the original species are
preserved. Pure-breed chickens are the most beautiful
option; they breed true and have distinct characteristics
which are recognised by official national organisations for
each breed. Hybrids are created by mating carefully selected
pairs of pure breeds, the cock of one breed with the hen of
another, and so on, to produce the best characteristics of both.

The array of breeds can seem daunting, although all
chickens originate from the jungle fowl of Asia. The breeds
can be broken down into two distinct groups: large fowl and
bantams. Bantams are the smaller birds and, as one might
expect, lay smaller eggs. Some varieties make wonderfully
docile pets and are very good for small spaces. Miniature
versions of many of the large breeds have been created,
however these are not true bantams.

Large breeds can be further sub-divided into Heavy and Light groups. The Heavies are both good table birds and layers – dual-purpose birds. They are quieter and more docile, but can get broody in the spring, although they lay earlier in the year. The Lights are excellent layers and consume less food, but they are nervous and flightier and far less inclined to be broody. Fancy breeds are the most exotic, a mass of fussy plumage, deliciously ornamental, but not of much practical use.

Each pure breed has its own set of technical specifications and these may vary slightly from country to country. In addition, each breed comes in many colourways and has varied markings to choose from. The technical terms seem complex and the novice may well find it difficult to understand the intrinsic differences between, for example, barring and striping (see Glossary). Seek advice from a professional breeder who will be delighted to explain the finer points of the breed to you.

ANCONA

The Ancona is a stylish monochrome bird; its looks change minutely so that they get gradually whiter with each moult – an ageing concept we are familiar with! This is a very robust bird which originates from Italy.

SIZE Light.

COLOUR Principally black and white mottled, with yellow legs and white ears.

EGGS 200 white eggs per year.

TEMPERAMENT This is a light and active bird, it is a good flier and will need a high garden fence to keep it in check. It is not very biddable as a pet!

Araucana

These birds are more famous for their blue or green eggs than their looks, in America they are known as Easter Egg Layers for their propensity to lay eggs in a range of shades. The breed appears both with and without a tail and usually carries tufts of feathers at the ear. The Araucana originates from South America.

SIZE Light.

COLOUR There are many coloured variants and international standards for the breed vary, most common are fawn, black and blue. Legs are grey blue.

EGGS Around 150 blue or green eggs per year.

TEMPERAMENT Araucana are strong and easy-going birds and cope with life in the run or the garden.

Brahma

Barnevelder

The Barnevelder is best known for her cloak of glossy feathers. She has a beautiful red comb, a red eye mask (face) and yellow legs and originates from The Netherlands.

SIZE Heavy.

COLOUR Beautiful warm brown and black tipped feathers, but she also comes in black, white and blue with beautiful double laced markings.

EGGS This breed lays around 200 dark brown eggs per year.

TEMPERAMENT Easy-going and friendly free-range chickens, inclined to become somewhat lazy if penned up in a run Barnvelders are not good at flying so garden fences contain them easily enough.

Brahma

These are seriously big chickens, a mass of fluffy feathers right down to their flared feathered trousers. The breed originates from India.

SIZE Heavy.

COLOUR They come in a range of colours and markings.

EGGS They lay around 150 cream eggs per year.

TEMPERAMENT Brahmas make good garden pets; they are relatively quiet, extremely friendly, easy to handle and make excellent mothers, but because of their size they do need a reasonable amount of room in which to roam.

BRAKEL

The Brakel is a truly handsome chicken with beautiful, glossy, sweeping neck feathers (cape) and contrasting mottled body feathers teamed with grey blue stockings. The cocks are very impressive creatures. The breed originates from Belgium.

SIZE Light.

COLOUR Golden or red brown and black feathers or blue, black and white feathers predominate.

EGGS A Brakel hen will produce around 180 gleaming white eggs per year.

TEMPERAMENT These are busy hens who appreciate a good deal of space and they are good at flying – so may not be an ideal urban chicken. They are independent and shy and not overly friendly.

FAVEROLLE

This is an exotic looking breed with a full beard, muffs, feathered legs and a fifth toe. The Faverolle originates from France.

SIZE Heavy.

COLOUR International breed variations occur, principal colours are salmon, white, black and blue. Feathers are often laced with white.

EGGS They are technically table birds, delicious to eat, however they do lay reasonably during spring and summer, producing around 180 light brown eggs per year.

TEMPERAMENT Faverolles are beautiful birds, very friendly and affectionate, and easily contained in a garden.

HAMBURG

The Hamburg is a very good-looking bird with beautiful markings and blue stockings. It originates from the Netherlands, Germany and the United Kingdom.

SIZE Light.

COLOUR Black, gold and silver spangled feathers and gold and silver pencilled markings are the best known varieties.

EGGS Hamburgs lay around 200 white eggs per year

TEMPERAMENT The downside to this stunning bird is that it needs plenty of space to roam in and, if limited to a run, tends to become aggressive. Definitely not suitable for small gardens.

LEGHORN

The Leghorn has many different varieties, and breed specifics change slightly from country to country. It is a very beautiful and popular show bird with exquisite spangled markings. It originates from Italy.

SIZE Light.

COLOUR Golden black or silver black spangles are most popular, but pencilled markings in yellow, gold and white also appear.

EGGS The Leghorn is a good layer, producing around 200 white eggs per year.

TEMPERAMENT They are happy in gardens with enough room to suit their active nature, but are shy and difficult to get close to.

MARAN

Marans are known for their pretty brown eggs and their strength and vigour. Some varieties wear feathered shin pads. The breed originates from France.

SIZE Heavy.

COLOUR Beautiful feathers in a striking variety of colours from warm red and black feather mixes to silver, copper and black combinations, some with cuckoo markings.

EGGS The Maran lays around 200 splendid brown eggs a year.

TEMPERAMENT The Maran is an amiable and easy-going hen well suited to the back garden, however she does not appreciate being touched or handled.

ORPINGTON

This bird is a real beauty, a mass of soft fluffy feathers with a personality to match. It is also very hardy and is a dual-purpose breed, being good for the table as well as laying. The bird originates from Orpington in Kent in the mid-19th century.

SIZE Heavy.

COLOUR This breed comes in buff, black, blue and white varieties with mottled and laced markings.

EGGS They lay around 180 eggs per year.

TEMPERAMENT Orpingtons make fabulous pets because they are friendly and gentle. They have small wings which, in relation to their overall size, make them poor at flying, so they are very safe for the garden. They are inclined to get broody in the summer and make excellent mothers, so if you want the full chicken and egg experience these are reliable performers.

PLYMOUTH ROCK

Plymouth Rocks have gorgeous yellow legs and red masks (faces). The breed originates from America.

SIZE Heavy.

COLOUR Plymouth Rocks are often very prettily striped, but also come in white, black and buff.

EGGS Produces around 200 cream eggs per year.

TEMPERAMENT This is a big, friendly, robust bird that is happier not penned in a run, although the average garden will be perfectly adequate for two hens.

Orpington

RHODE ISLAND RED

This bird is a favourite worldwide for its pleasant nature, its strength and because it is both a good layer and a delicious table bird. Rhode Islands are not skilled fliers because of their size. The Rhode Island Red originates from the USA.

SIZE Heavy.

COLOUR It is best known for its warm brown feathers with a dark head and tail tip.

EGGS These friendly birds lay around 240 brown eggs per year and because of their laying capacity are frequently used to create many of the hybrid hens used by commercial farms.

TEMPERAMENT These are amiable, easy-going birds. They are a pleasant bird for a beginner.

Rhode Island Red

Sussex

The Sussex is a really pretty and gentle hen. She is a dual-purpose bird – both a good table bird and a fantastic layer – and this is a great bird for beginners. The breed originates in the United Kingdom.

SIZE Heavy.

COLOUR Best known for their white bodies, black-laced necks and a sweep of black tail feathers, but also come in silver, buff, red and brown.

EGGS The Sussex, like the Rhode Island Red, is a strong layer and can notch up 260 eggs per year. Eggs range from cream to light brown in colour.

TEMPERAMENT A calm and friendly garden bird, one of the most amiable to keep.

Welsummer

Welsummers are famous for their large, speckled, terracotta-coloured eggs. The breed originates from the Netherlands.

SIZE Heavy.

COLOUR The feathers are warm russet, chestnut and black, in stripes on the hen. The combs are small and the legs yellow.

EGGS They lay around 200 beautiful brown eggs a year.

TEMPERAMENT These are easy-going birds that enjoy rummaging through the garden for their food.

BANTAMS

JAPANESE BANTAM

Japanese Bantams look like an Ascot hat on exceedingly short legs. They are ornamental birds that, because of their short stature, demand special facilities – low perches in their house and a low door as well as a dry run in wet weather. They come with beards, purple combs and frizzled or silk feathers. The Japanese Bantam, you will not be surprised to learn, originates from Japan.

SIZE Bantam.

COLOUR: These come in a huge variety of colours, feather types and markings, including black, blue, white, yellow, wheaten colourways and mottled, pencilled and partridge markings.

EGGS They are not good layers and their very few white eggs are tiny.

TEMPERAMENT They are sweet and pretty birds that don't do much damage in the garden. They are very gentle and good with children although the cockerels can have their macho moments! They do well in courtyard gardens with less soil and wet foliage for them to drag their feathers through.

OLD ENGLISH GAME BANTAM

The original fighting cock. This is a small, muscular breed which originates from the United Kingdom.

SIZE Bantam.

COLOUR Typically appears in shades of partridge, black with red and gold detailing; but also available in a huge array of colours and markings.

EGG These birds are not good layers; they produce only a small number of eggs in spring and summer.

TEMPERAMENT This breed is incredibly quiet and friendly and really good with children. The hens make excellent mothers.

PEKIN BANTAM

The Pekin boasts gloriously long, soft feathers, which it wears right down to its feet, making it look as though it is dressed in flares. The breed originates in China.

SIZE Light.

COLOUR It comes in many colours: white, lavender, buff, blue and black with mottled and cuckoo markings.

EGGS The Pekin lays only 60 beige eggs per year.

TEMPERAMENT The Pekin is a superb character for a small garden. It is good with children and appreciates having some grass to peck at.

SILKIE BANTAM

A flamboyant but diminutive bird with a luxuriant mass of soft feathers and an exotic pom-pom crest hairdo. Technically this bird is not a bantam but, because of its size, it is often grouped under this heading. The breed originates from China.

SIZE Light.

COLOUR Black, blue, gold and grey colourways with pencilled and laced markings. The plain white bird is deeply glamorous.

EGGS The Silkie lays around 100 beige eggs per year.

TEMPERAMENT These are gentle birds that make good garden pets although because of their fluffiness they do need a dry run. They are perfect for very small gardens. A friend has a Silkie that will roll over and play dead on command! They are broody and make excellent mothers, regularly producing batches of chicks given the right conditions.

WYANDOTTE

The Wyandotte is a beautiful bird it is actually a miniature breed rather than a true bantam. It has spectacular markings, a bright red comb and eye mask and yellow legs. The tail fans out in an inverted V shape. The breed originates from the USA.

SIZE Miniature (also available as a Large).

COLOUR This bird is available in a wonderful range of colours including white, black, blue and buff with laced, pencilled and partridge markings.

EGGS This bird lays around 150 light brown eggs per year.

TEMPERAMENT Wyandottes are busy birds, but are also gentle and friendly, easy to tame and make good garden pets.

Barnevelder

Ancona

Hamburg

Leghorn

Plymouth Rock

Pekin Bantam

Sussex

Japanese Bantam

Brakel

Araucana

Old English Game Bantam

Maran

Faverolle

Wyandotte Bantam

Silkie Bantam

HYBRIDS

BLACK ROCK

The Black Rock hybrid is a spectacularly good layer – I know, as I have one! They make a resilient garden hen, and are the product of a cross breed of Rhode Island Red and Plymouth Rock

SIZE Medium.

COLOUR A combination of glossy russet and black feathers.

EGGS Black Rocks lay around 300 beige eggs a year! Mine, in her prime, laid even more!

TEMPERAMENT The birds are friendly enough, extremely curious and will come when their name is called (most of the time). They don't like being handled. If you don't clip their wings annually they will explore neighbouring gardens!

SPECKLEDY

This hybrid, a Maran and Rhode Island Red cross, is a sweet chicken with a slightly glamorous look.

SIZE Large.

COLOUR This is a very pretty chicken with grey, brown and black feathers, a bright red comb and a red eye mask to match.

EGGS Up to 250 dark brown eggs per year.

TEMPERAMENT This is a mild and friendly garden bird. It doesn't enjoy being handled, but is easy enough to catch.

Battery hens are in peak egg-laying prime up to a little over one year old; after that they are of no use to the farmer and will be discarded in favour of younger models. They can be obtained via rescue charities and will inevitably be hybrids, farmed for their egg-laying capacities. If you are none too fussy about your breeds and simply want to help out some poor old bird that has had a miserable existence then these birds are for you. They will look terrible on arrival – bald and fearful. But give them a little time, space and tlc and they will pluck up the courage to explore the outside world and eventually acquire a fine set of feathers. They will also continue to lay, perhaps just a little less frenetically, but lay they will. My Black Rock Hybrid, Wilma, is four years old now and managed to lay a good 180 eggs last year.

BUYING A BIRD

Your initial decision should revolve around what breed to buy. Once this decision is made you can source a supplier. Poultry clubs and organisations will have lists of nationally recognised breeders and breed clubs who can be contacted via websites, as will specialist magazines and local newspapers. You may purchase a bird, but have to wait patiently for several weeks until it is big enough to leave home. For some unusual or rare breeds this wait can be as long as six months while your order is fertilised, hatched and reared to point of lay. Sometimes there is a boom in sales and breeders simply cannot supply demand. More usually, you will be able to source suitable point of lay birds within a month or six weeks.

You can buy your birds at any stage in the growing

process. However, for a beginner, the simplest and easiest place to start is with the purchase of a point of lay (POL) pullet – a female bird that has yet to lay an egg. They usually start at around 18–24 weeks old. Their combs may be small and pink rather than large and red as they will not yet be fully developed. A fine red comb is indicative that the bird is about to start laying.

Technically a bird who has been through an annual moult could be described as point of lay – but you should not buy a bird that is more than one a year old as its best laying period is already over. Birds are often advertised for sale as a trio – this means two hens and one cockerel so do not assume that you will get three hens!

You may want to mix two different pure breeds together, but it is simplest to get two or three hens that have been reared together, this way they won't fight. Once you get used to keeping chickens you will have more confidence about introducing a newcomer – see A Year in the Life of a Chicken Run. Don't ever buy just one bird: chickens are sociable flock birds and they become very depressed if they are alone. Two birds or more is the simple rule – your girls will quickly become inseparable.

If you do not want to have a cockerel and, beautiful though they are, this is definitely the simplest option, make sure that there is no confusion regarding the sex of the birds you require. A cock will be vocal first thing in the morning, he will rule the roost very firmly and your hens will be rather more inclined to broodiness with him in the run.

When you buy, check that the birds seem healthy, that the feathers and eyes are bright and that the bird is active. You can ask to give the bird a once over and check under the feathers and around the vent for parasites. A reputable breeder will want to sell you a strong and healthy bird and will be pleased to show their birds off and explain the characteristics of the breed.

Prices vary depending on the breed you want to purchase. It is a little shocking that purchasing a hybrid, the cheaper option, will generally cost considerably less than the bird you find trussed up on the supermarket shelf ready to cook. Pure breeds will cost more and the price will vary according to the availability and quality of the breed.

If you want to hatch eggs yourself this is perfectly possible and, in terms of purchase price, it will be considerably cheaper. However, you will also have to buy specialist equipment. You will need an incubator, a brooder or, much simpler, an obliging broody hen that you can borrow to do the job for you. A broody hen is the perfect solution, allowing you to have the full chicken to egg experience first hand, with none of the responsibility – the broody will take care of the chicks. Please remember that, although this is a wonderful

Welsummer

experience, the drawback is that it will be impossible to predict what your eggs will produce – cockerels or hens – so you will need to decide how you will deal with any unwanted birds. See A Year in the Life of a Chicken Run for more information on raising chicks.

EGGS

Eggs are a miracle of construction, one of the most extraordinary pieces of design on the planet, both sophisticated and practical. Indeed, if you have ever been to a village egg-throwing competition (great fun, keen cricketers almost inevitably win), you will know that they are also remarkably resilient, even though an eggshell is only around 0.3mm thick.

HENS WILL LAY whether or not their egg has been fertilised, so you do not need a cock to have a fabulous supply of eggs. Fertilised eggs can be eaten as long as the hen has not sat on them. There is no apparent difference in taste between fertilised and unfertilised eggs. Fertilised eggs should, however, be eaten within two weeks of being collected.

CONSTRUCTION OF AN EGG

The ovary releases an ovum, or yolk on a regular cycle. The yolks are contained in a cluster in which they gradually ripen over a period of several days, until the one that is ripest passes down into the oviduct for the next stage of production. It is at this stage in the process that the egg would be fertilised. It rotates along down the oviduct and, as it does so, the walls coat it in a thick layer of albumen or egg white. Next, a thickened cord called a chalaza is added to each side (those stringy bits in the egg white), these act as buffers to support the yolk.

The final stage within the oviduct involves two membranes encasing the egg. These inner and outer membranes separate and, at the larger end of the egg, form the air space, which is small in new-laid eggs and increases in size according to the staleness of the egg – which is why rotten eggs float in water.

It takes approximately 20 hours for an egg to travel down the oviduct and there are usually a few eggs within it at any one time, in various stages of production. In the final chamber of the oviduct the whole thing is coated with a thin liquid secretion of lime, which quickly hardens to form the shell. It then passes through into the cloaca and proceeds outwards via the vent.

Blood spots occur due to a rupture of a blood vessel in the ovary, they are most common in the yolk. If your chicken is frightened this could also cause a blood spot – for instance, chicken keepers often find many more bloodspots in their eggs after a thunderstorm. If you find a blood spot on cracking an egg, be reassured that it is perfectly safe to eat – although you may prefer to pull it out using a teaspoon.

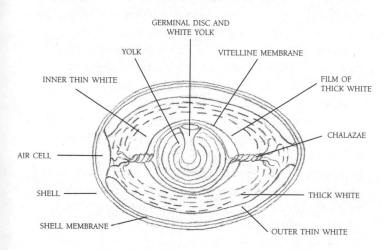

The egg: a miraculous design

Eggs can make a significant contribution to a healthy diet. They are packed with nutrients, are relatively low in saturated fat and are low in calories – only 78 calories per medium egg. Of the egg content 10.8% is fat. The fat of an egg is found almost entirely in the yolk, there is less than 0.05% in the albumen. Approximately 11% of an egg's fatty acids are polyunsaturated, 44% are monounsaturated and only 29% saturated.

According to The Royal Society of Chemistry MAFF 1991 *The Composition of Foods* (5th Edition), eggs are a good source of protein of high biological value (the yardstick by which other proteins are measured), as they contain all the essential amino acids needed by the human body. A medium egg provides 12% of a man's and 14% of a woman's daily protein requirement and it is found in both the yolk and the white of an egg. Eggs contain most of the recognised vitamins (with the exception of vitamin C): A, D, E, B6, B12, thiamin (B1), riboflavin (B2), niacin, folate, biotin and pantothenic acid. Vitamin E gives protection against heart disease and some cancers, vitamin D is also present to provide mineral absorption and good bone health.

Eggs contain antioxidants such as selenium. They are rich in essential minerals, notably iodine, which is needed to produce thyroid hormones, as well as phosphorous for healthy teeth and bones. In addition eggs are also a great source of choline. Research has shown that it is essential that women have sufficient choline during pregnancy, as without it the baby's brain doesn't develop normally and it can be born with either defective memory or lower memory capabilities that last throughout life.

Eggs provide zinc, for improved immunity and wound healing, and calcium, needed for bone and growth structure and nervous function. They also contain quantities of iron, however it is uncertain how much of this iron is available for take-up by the body

It used to be believed that consuming food containing dietary cholesterol, such as eggs, butter, cream, liver, kidney and shellfish, directly increased cholesterol in the blood. Anyone with high cholesterol was advised to limit his or her intake of eggs to one per week. However, it is now believed that eating too much saturated fat, which eggs are low in, causes high cholesterol.

The Department of Health in the UK recommends that vulnerable groups such as the very young, pregnant women, the sick or the elderly must take care that the eggs they consume are thoroughly cooked – i.e. no mayonnaise made with raw egg yolk!

ALLERGIES

The British Egg Information Service highlights that egg allergy is most common in infants under the age of 12 months, after which it generally becomes less of a problem. Since eggs are often introduced early in a baby's life, they can cause problems, however few children are allergic to eggs after the age of six.

Some foods are more liable than others to provoke allergic reactions because of the proteins they contain. Some proteins are more digestible, stable and absorbable than others. Eggs contain proteins that, in their raw state, are of the right size and stability to cause allergies but which, in most cases, cannot withstand the effect of heat. This explains why people with a mild allergy to eggs can often tolerate them in cooked food such as pasta or cakes.

More than half of the infants who develop an egg allergy begin to have symptoms within minutes of being given an egg for the first time. Whilst it is possible that some received small amounts of egg in a manufactured baby food, it is also possible that some have been sensitised before birth or via breast milk. The allergy presents itself as a red rash around the mouth, followed by swelling inside and around the mouth and face. Later further areas of skin swelling or eczema can occur, together with sneezing, wheezing and runny eyes. If a child develops local swelling on contact with raw or cooked egg this a strong indicator of a likely problem.

Allergies developing in adulthood are far less common, though the symptoms are similar to those experienced by children. It is important to seek the advice of a dietician if you or your child develops an allergy, as eggs are contained in all sorts of unlikely foods.

NUMBER OF EGGS LAID

The number of eggs a bird will lay varies from breed to breed. Hybrids have been bred to lay around 300 eggs per year, though this constant laying does wear the birds out and they die younger than pure breeds, If it is eggs you want (and this is the main fringe benefit of keeping chickens), some of the pure breeds, such as the Rhode Island Red and the Sussex, lay almost as well, producing as many as 260 eggs per year.

In the United Kingdom some 28 million eggs are consumed daily, most are purchased from supermarkets and approximately 30% of the market is free-range. Once you've tasted an egg from your own chicken you will never want to buy a shop-bought egg again – not even a free range one!

It is a common misconception that free-range eggs are brown, ie., that it is the free-range lifestyle that makes hens lay brown eggs. In reality, the breed of the hen determines the colour of the egg and you can also get an indication by looking at the colour of a hen's ear; birds with white ears lay white eggs and birds with red ears lay brown eggs. It does, however, go some way toward explaining why every egg on offer in the supermarket today is brown, free range or not. Brown eggs are, incorrectly, regarded as healthier eggs – even if the birds that laid them are penned up in a tiny cage. White eggs have become something of a rarity today; in fact when I next need to replace a chicken I am determined to select a breed that lays beautiful white eggs.

Egg colours range from white through beige and on to warm terracotta browns and can come in pale blue or pale

Home is where the eggs are laid

green – these are laid by the Araucana.

The colour of the egg yolk is determined by a bird's diet. A bird that is getting sufficient greens, whether through grass, salad leftovers, broccoli or spinach, will lay eggs with a vivid orange yolk – everything you cook with them looks better; sponge cakes are a warm gold, scrambled eggs are canary yellow and a Spanish omelette a noisy mix of bright yellows, reds and greens.

UNUSUAL EGGS

Double-yolked eggs occur when two eggs separate from the ovary at the same time and join together. Or a yolk may not proceed smoothly as it moves through the production process, not finding its way into the oviduct's entrance but slipping into the body cavity for a while before rejoining the oviduct, where it joins another yolk as it is moving along. Double yolkers occur with more frequency in young hens. Similarly if your bird lays a tiny egg you will find, when you crack it, that it does not contain any yolk at all, again a fault in the production process. Both of these problems are more likely to occur in very young or very old birds.

SHAPES AND SIZES

Eggs are sold in four different sizes: small, medium, large and very large.

Small eggs weight 53 grammes and under, medium eggs weigh 53–63 grammes, large eggs weight 63–73 grammes and very large eggs weigh over 73 grammes.

There is no connection between the size of a bird and the size of eggs she lays, although bantam's eggs are generally the smallest. However, the fluffy Brahma, a large heavy bird, also lays very small eggs. Scientists from Oxford and Sheffield University have discovered that cocks are more attracted to hens with large fleshy combs or crests because they are healthier and lay larger eggs.

SHELF LIFE

EU legislation stipulates that the maximum 'best before' date must be no more than 28 days from point of lay. Some nationally recognised quality marks, such as the Lion Mark in the UK, insist that the best before date is shorter – 21 days from point of lay. Shop-bought eggs are generally packed within 24 hours of being laid and are then dispatched to the shops. Therefore, the very freshest off the shelf egg that you are likely to find is two days old, but that would be an unusually fast turn around.

Eggs marked free range are from birds that have been allowed to roam freely in a farmyard and are only penned in sheds or henhouses at night. Free-range eggs are said to contain higher levels of Omega 3 oils.

If you want to date your eggs then write the date on them with pencil, egg shells are porous. Though if you have just a couple of chickens you are likely to find that you consume most of the eggs as they are laid. Always wash your hands before collecting and store the eggs in a clean container. Dirty eggs should be cleaned gently with a soft damp cloth before putting them with other eggs. However, ideally your eggs will be delivered to you nice and clean and will not need wiping. An egg shell is porous, but coated in a protective bloom that will not allow anything to pass through into the egg – when this is washed away the egg has lost its natural defence and can be penetrated by bacteria – which explains why poultry farmers are not allowed to sell eggs that have been contaminated with droppings.

It is important to collect eggs daily. Not only does this ensure that you always know which are the newest, but it also discourages chickens from breaking and eating eggs. See the chapter on Problems for more information.

If you uncover a pile of eggs in a secret garden nest and you are unsure of their vintage and whether they are safe to eat, you can easily test them. Fill a bowl with water and place the eggs in it. If any float, they have gone bad. If they sink, they are fresh. Similarly if an egg stands on end, it will be just a few days old: either way you can safely consume it.

Eggs can be frozen, though they need to be separated first. Freeze the whites in quantities of two store the yolks by first mixing with either half a teaspoon of salt or half a teaspoon of sugar, for each three yolks. Once separated, eggs can be stored safely in freezer bags for up to three months.

DROP IN EGG PRODUCTION

Chickens naturally lay fewer eggs in the winter as they are light sensitive and therefore their productivity is affected by the shorter daylight hours. The most common reason for a chicken to stop laying eggs over several weeks is that she is moulting, when all her energy goes into feather production. You can't miss this when it happens as the chicken house and run fills up with feathers and the bird looks somewhat bald and very sorry for herself.

When a bird has been chased by a dog or escaped from a fox, she will probably also stop laying for a while. Fright affects egg production, and you are advised to bring your chickens into your house on 5 November (in a box). Lack of water can also affect your bird's ability to lay, so you must always ensure that there's a plentiful supply available and check twice daily on very hot days as your girls may want a

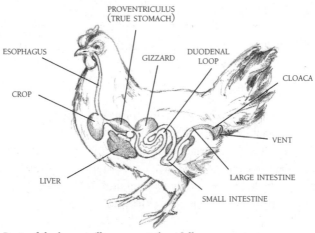

PROVENTRICULUS
(TRUE STOMACH)

ESOPHAGUS

GIZZARD

DUODENAL
LOOP

CLOACA

CROP

VENT

LIVER

LARGE INTESTINE

SMALL INTESTINE

Parts of the hen you'll never see - hopefully

good drink of water before they retire for the evening. If you have concerns you will find more information in the chapter on Problems.

WEIRD SCIENCE

Scientists have turned their attention to the medical possibilities of using eggs to fight cancer and other life-threatening diseases. A flock of genetically modified ISA Browns (a cross between Rhode Island Red and Rhode Island White chickens) have had human genes added to their DNA to enable them to produce complex medicinal proteins. Human DNA is injected into embryo cockerels and these GM cockerels are then mated with ordinary birds. The resulting offspring have human genes bred into them. Human proteins are secreted into the whites of the birds' eggs, from which they can be extracted to produce drugs.

COOKING PERFECT EGGS

Fresh eggs do not need to be refrigerated, though they should be kept in a cool place.

BOILED EGGS
First fill a pan with water and bring to the boil. Eggs stored at room temperature will not crack when put into boiling water. If they have been refrigerated, you should prick a hole in one end of the egg with a pin to prevent this happening. A large egg is soft boiled in four and a half minutes. Try dipping asparagus tips into the yolk instead of bread soldiers.

SCRAMBLED EGGS

The secret to good scrambled eggs is to get everything ready and then concentrate on cooking them. Melt a knob of butter gently in a pan. Do not let it bubble or brown. Meanwhile whisk the eggs well, using two to three per person. If you have plenty of eggs, add a couple of extra egg yolks to make the dish even more delicious. Season with salt and pepper and mix well again. Place the saucepan back on a gentle heat and coat it with the melted butter, add the egg and keep stirring with a wooden spoon. Scrambled eggs suddenly accelerate towards a perfectly formed, creamy consistency, in a last-minute rush, so keep taking the eggs off the heat to control the rate of cooking. As a special treat try adding a little cream, or some chopped smoked salmon before cooking the eggs.

BAKED EGGS

My children adore this when they're feeling a little fragile. Heat the oven to 180°C/ 350°F/ Gas Mark 4. Melt some butter in a pan and then use it to grease a cocotte dish. Break an egg into the dish, season with salt, then pour on cream, add a touch of butter to the top. Place the cocotte dish into a bain marie and bake in the oven for 12 minutes.

HEN HOUSES AND RUNS

In theory, chickens do not need specialist housing as they are perfectly capable of looking after themselves; foraging for food, roosting in trees and shrubs and making nests in secluded, shady corners. Some of the Greek Islands are alive with hens. They roost all over the place, wandering and breeding as and where they please. In fact, the only apparent clue to any human input is some little ladders running from the ground up to the branches of the olive trees, enabling the birds to roost in safety.

IN MOST COUNTRIES hens need housing to protect them from the elements. Although chickens are quite hardy they do not like being muddy or wet and they hate wind – clearly Greek winters are not as wet and windy as those in northern Europe. Last, but not least, they need protection from predators, principally foxes, though rats, dogs, badgers and mink can also cause problems. It's also more convenient if they lay their eggs in the hen house.

The threat from foxes must not be underestimated and, put simply, if foxes regularly visit your garden, it will be impractical for you to keep hens. Given the opportunity, a fox will kill all your birds even if it only takes away one to eat. A fox is also determined, and will work on catches and door locks, and will attempt to tunnel under runs. So all chicken housing and runs must be well designed and robust, with secure locks. Similarly, any doors for egg collection must be secure. Mice and rats must also be kept out of the chicken house. For more information on predators see the chapter on Problems.

Chickens have specific requirements if they are to be housed comfortably, but a chicken house can be made out of anything that will allow you to incorporate the essential elements: pop hole (to allow chickens in and out of the house), roosting bars, accessible nesting boxes (one for every three birds), good ventilation to help remove moisture and to prevent build up of ammonia from droppings. Good insulation will keep the chickens warm and draught free in winter, cool in the summer. Last but not least, the house must have surfaces that can be cleaned easily. Good hygiene will help to keep your chickens healthy and happy. Chickens can be kept in modified garden sheds, unused garages, caravans and log stores.

Purpose-designed chicken houses are not cheap – though the range of prices is somewhat dependent on how functional or ornamental a chicken house you require. If you want a domed cupola with bell tower on your chicken house

The hen house
is a place of safety

someone will make it for you a price. What purpose-built houses offer is security, easy access for cleaning and portability. Even if you only have two hens in your garden you will probably need to move your chicken house regularly – if only to give your birds a fresh supply of grass to eat, although your lawn will also thank you for repeatedly moving the hen house. I cannot emphasise the importance of portability and easy access for cleaning – keeping chickens should be a positive pleasure, not a nightmarish chore.

If you are lucky, you will have the luxury of being able to fence off a section of your plot and thus ensure that your birds can only make their presence felt in one area alone. I have fantasies about owning an orchard and having 12 pure breed chickens scratching about in it. Be aware that if you are keeping hens free-range, you are not allowed to have more than 1000 birds to an acre.

For the time being however, I, like many other people, am limited to the two or three birds I can accommodate in my back garden. Two chickens need about 50sq metres minimum to roam in. It is important that you choose a good site. It should be well drained as chickens cannot cope with boggy ground. In winter, chicken houses should face south east, to maximise light levels and warm up the house. If you are subject to strong prevailing winds make sure that the back of the house faces the wind to minimise draughts.

Any house must be wind and water resistant and, while well ventilated, it must not allow temperatures to drop too much. If the house temperature drops below -12°C, egg production will drop by 25%; if it drops below -17°C, then no eggs will be laid at all. Similarly, birds become unhappy when temperatures rise above 26°C. Egg production will be affected, especially if temperatures do not drop noticeably at night.

Hen houses must be roomy and robust

In terms of the size of the house, you should allow a minimum of 30sq cm per bird, more if you are housing heavy breeds. This means if you are just keeping two or three chickens, or four to five bantams, that your housing will be quite small, approximately a metre squared and can easily be accommodated in an average back garden. Manufacturers will specify how many birds a house can accommodate, and it is important that this guidance is respected. A house may apparently be able to contain another one or two chickens in terms of space, but ventilation would be compromised. A couple of bantams can be housed in a customised rabbit hutch (as long as it is on legs).

If you are a DIY enthusiast and can build your own, or if you are purchasing ready made, you need to keep certain key features in mind.

Chickens generally prefer to sleep at night on roosting bars or perches. Their feet are quite large and to grip comfortably this should be approximately 25–35mm wide with curved edges. The size of the roosting bars is also affected by the breed you want. Round roosting bars are not recommended as it is hard for the birds to grip effectively and, similarly, the broadest side of the perch must be horizontal. Roosting bars must be rubbed smooth, there must be no splinters or jagged edges that can injure the birds, or cracks and crevices which will house lice and mites, and they should be designed to lift out easily for cleaning.

All perches should be at the same height, otherwise the birds will jostle to be on the highest perch. However, heights vary depending on the breed you wish to house: bantams need perches around 15cm from the ground, whilst the heavy breeds can cope with roosting bars no more than 30cm from the floor – these outsize birds must be able to hop on and off with ease. Birds like to huddle together for warmth on their perches and by doing this and by crouching, so that their legs and feet are protected by their feathers, the birds lose very little body heat. Chicken houses should not be heated – otherwise the birds will be shocked by the cold when they emerge in the morning, and are likely to fall ill.

Birds are supposed to do 50% of their mess at night, therefore a droppings tray or board should be placed underneath the perches to catch mess – and, depending on your enthusiasm, this can be emptied daily or twice weekly to keep the place smelling fresh. It also makes the weekly clean up very easy.

On emerging in the morning, my birds always give themselves a good shake before producing a very large dropping in the middle of the lawn!

NESTING BOX

Chickens take anything from an hour to two hours to lay their eggs and they must have somewhere comfortable to do so. Indeed, if they are unhappy with the one you have provided, free-range birds will find their own nest. Nest boxes are enclosed – except obviously at the front and the top – and they should be shady and private. Darkness also discourages birds from eating their eggs. If this is a problem, get a nest box with a double bottom, which directs the egg out of the nest box and into a soft lined chamber where it is held safely until it can be collected.

Some houses have a dish for grit sited near the nest so that the chicken is able to peck away as she lays, others put a layer of grit in the nest box itself. It is also nice for the hens to have a good lining of straw in their nesting boxes; they like to tidy this, fussing with their beaks as they cogitate. Straw in nesting boxes should be changed regularly to avoid the eggs being contaminated with droppings. I find once a week is sufficient in summer but twice a week in winter when the birds are in the house for longer. Hay is not recommended.

If you have more than one nest box make sure that they are sited on the same level. The same rule applies as for perches: chickens will fight for the best vantage point.

Pop holes

Chickens use the pop hole to move in and out of their house and often to access their run. The pop hole must be large enough to accommodate their comings and goings in comfort – average dimensions are 40cm high by 30cm wide. The doorway must be nice and smooth so that birds cannot accidentally injure themselves. Leaving the house in the morning is often a bit of a free for all, and birds can easily get cut if edges are not well sanded. Door designs vary but anything goes as long as it can be locked securely. If you don't want to bother with locking every evening and unlocking in the morning (though personally I enjoy it), timer operated door systems are available. If your door is solid (not partially meshed) it will help minimise draughts and light – the latter an important consideration if you have a cockerel and want to avoid upsetting the neighbours.

Maintenance

The floor of the house should be sprinkled with straw, sand or wood shavings and this should be changed weekly. If you have a wooden chicken house, you will need to treat or paint the wood annually to ensure it stays in good condition. You do not need to provide food and water within the chicken house, your birds will be very happy to feed in the hours of daylight in their run and they usually have a bit of a drink and a nosh before they retire for the evening. Similarly, if your run is roofed and space permits, you could try making a roosting swing for your chickens, they seem to quite enjoy it.

RUNS

A run will allow your chickens access to the outside world in complete safety. If you have just a few birds, a run attached to their house is often a good option; you can let them out into this in the morning and either leave them there or open one end and let them loose in the garden. My chickens roam free for nine months of the year and the fox leaves them alone. But from January to the end of March, when food is scarce, they are firmly locked away in their run as the fox will brave the garden even in daylight. I have lost two chickens to the fox and it is very upsetting. Moreover, you then have to go through the palaver of settling in a new chicken. So for three months of the year my girls have their right to roam curbed, although we still try to give them a little time out every day, but we make sure one of us is always out there with them.

A run should give each bird a minimum of one square metre in space. It will need to be sturdy and roofed. Although many breeds cannot fly more than a few feet high, the fox can climb beautifully and I have seen one jump over a 1.8 metre fence, so don't rely on a tall roofless run keeping the fox out.

The cuddly fox thrives on chicken meat

A roofed run does not need to be tall, if it is just 60cms high it will be sufficient.

You will also need to think about the floor of the run. It is easier to clean if the floor of the run is not meshed. If the fox is determined to get at your chickens, he will simply dig underneath. If your chicken run is in a permanent site you can put posts and wire deep into the ground as a deterrent, or on smaller runs you can make a skirt of wire which extends around the outside of the run. Runs can be made of conventional chicken wire or welded mesh for greater strength.

If you are fencing off a section of your garden for the chickens' exclusive use, you can always opt for electric fencing – this is costly, but effective, as long as you always remember to turn it on. Ringing the whole garden with electric fencing can be prohibitively expensive, and it won't work if you have any form of gated access to the garden – a fox will find the weak point and use it.

If chickens are kept in a run, part of it should offer shelter from the rain. Your birds must be able to find a dry place to peck about in, and it is a good idea if food is kept within this area so that it doesn't get spoiled every time the heavens open.

A run can be housed in a flower bed If you put bark chips down, it will not have to be moved so frequently; you can just rake up the old bark chips and replace with new when it looks the worse for wear. This also saves the lawn from wear and tear but won't work if your flowerbeds are not sufficiently roomy or flat enough.

You should always check with your local council and look at the deeds of your house in case you are prohibited from keeping livestock. In addition, if you are planning on keeping more than just a few birds and want to build a largish chicken house then you may need a building permit.

PROBLEMS
BEHAVIOUR

Chickens are not always the most obliging creatures and they have distinct character traits that can cause occasional problems.

BROODINESS

Chickens become broody when they sit on eggs in an attempt to hatch them. It is most likely to happen in late spring and summer. Some breeds are much broodier than others and they refuse to leave the nest and get very grumpy if you try to remove them. They will also resume position as soon as your back is turned. Birds sit on their eggs for 21 days – so if you allow your bird to attempt to hatch unfertilised eggs she will stay put for this period and probably longer – as nothing will happen! She will also inconvenience all your other birds who

Hatching a plot:
a broody hen on her nest

will be kept off the nest box. Collecting eggs promptly is one of the simplest ways to discourage broodiness. If you have a broody hen separate her from the other birds and put her in a special house, ideally with a slatted or wire floor and no nest box; she will find it difficult to squat and this will discourage her inclination to motherhood. It may be two weeks before she resumes laying.

BULLYING

The term 'pecking order' originates from chicken behaviour. Chickens quickly establish who is top chicken – the godfather if you like – and work down from there. If there is no cockerel in the flock a female will take on the role of leader. Newcomers are given a very hard time indeed as they are a potential threat to the status quo. Similarly, a sick or injured bird will not be treated with any compassion. The rest of the birds will ruthlessly exploit any weaknesses. Birds will pull out the feathers of another hen and attack her vent until it is sore and bloody. A bird that is being given a hard time should be separated from the rest of the birds and given her own food and water supply to try to improve her health. This is not necessarily as difficult as it sounds – a cardboard box/cat box in the house (somewhere cool) overnight is fine, and if you don't pen your birds full time then the weak chicken can be locked in the run with her food and drink while the others roam free. Alternatively, fence off a tiny section of the garden with chicken wire as a temporary daytime measure. Birds living in cramped or overcrowded conditions are more likely to resort to this kind of behaviour.

Egg eating

Eggs should always be collected promptly to prevent this problem occurring – though admittedly this is not always possible if you are out at work. It is also helpful if nesting boxes are dark. However once a chicken has broken an egg and eaten it she will probably do so again and again. A friend has informed me that you can try blowing an egg and then refilling it with mashed potato heavily laced with mustard and pepper – which the chicken will not like. Another option is to purchase a new nest box, one with a double bottom, which will direct the egg out of the nest box and into a soft lined chamber where she can't reach it, but you can collect it.

Escape

Hens can be very perverse and, even though they have the most wonderful space to roam in, they are likely to want to explore new territory if the opportunity presents itself. If your hen is annoying the neighbours with her constant visits, check the primary feathers on her wing – if it is post moult she will have grown fine new feathers and will need to have her wing clipped afresh, for more information see the chapter on A Year in the Life of a Chicken Run. If her wing is already clipped, then you will have to keep an eye on her to see where and how she is escaping – though hens seem to know when you are watching and act innocent until your back is turned. Wear and tear may have produced a weak spot in the fence that she can exploit. Once your birds find their escape route is plugged they will resign themselves to staying put.

Garden damage

Hens view your beautiful garden as one big larder, and they will be none too careful how they access the delicacies on offer. If you are a fastidious gardener, you would be well advised to invest in some low hurdle fencing, birch or hazel twigs for protective wigwams, cages, or even just a length of chicken wire and some bamboo canes, to protect new seedlings. The birds will be as much attracted by the finely raked expanse of soil, as by the tender young growth and will do far more damage having a good scratch around than by nipping the tops off your new shoots. Similarly, it is a good idea to put pieces of chicken wire over newly planted bulbs. Once they come through they will be safe but, until then, the birds will want to scratch around in the soil. Vegetable gardens will need fencing off if your chickens are not to scratch them to pieces.

Noise and smell

It is not unheard of for a cock's early morning crow to cause friction with the neighbours! The local council's environmental health department will deal with complaints about this as they would with any noise nuisance. If you live in an urban area it is probably simplest if you steer clear of male birds. If you have a broody hen you can still incubate chicks by buying fertilised eggs from a breeder. Do remember though, some of those dinky fluffy little chicks will inevitably grow up to be cocks and then you will have to face culling them. It is hard to find someone willing to take a cockerel off your hands!

If you just keep a few chickens, then your chicken house should not smell if you are cleaning it out on a regular weekly basis. Emptying the dropping tray daily helps minimise unpleasant odours. Difficulties are only likely to arise if you want to keep large numbers of chickens in a small space, and this is not recommended for their health.

HEALTH

If your chickens are well looked after, given the correct food and drink and their house kept clean, they are unlikely to have any significant health problems. However, it can be very worrying, especially when you are a novice chicken keeper, if a bird suddenly seems to be off colour. Wild birds can spread some complaints, and there is little you can do about this. The following list is by no means all encompassing, but it covers some of the most common problems. If in doubt you should always pack your chicken into a box and take her to the vet.

COLDS AND BIRD FLU

Chickens can catch cold. Symptoms include sneezing, slimy nostrils and eyes, and beaks hanging open – because the bird cannot breathe. Treatments are available from pet shops and vets to help the bird recover. It is a good idea to separate this bird from the others to stop the virus spreading. If your bird has caught a cold, her house may well be too chilly or damp.

While bird flu can hardly be classed as a common problem, it is so newsworthy that it has unnecessarily, but understandably, discouraged many people from taking up chicken keeping.

Bird flu is primarily a disease caused by flu viruses. There are many different viruses and each differs in its ability to cause disease in birds and people. People can become infected through close contact with diseased poultry. However, infection in people is rare, even in those who are directly exposed to the virus, because it does not pass easily from birds to people. It has been around for many years, and the first recorded scientific research was done around 100 years ago. To become a virus deadly to humans it must mutate into a human virus, which could then go on to cause a worldwide epidemic of flu – a pandemic. This last occurred in 1918–1919 and killed around 40 million people worldwide.

Bird flu is passed from bird to bird by direct contact with secretions and faeces, and through indirect contact via water bowls, feeders and boots. Symptoms include: loss of appetite, breathlessness, blue discoloration of the comb and wattle, oedema (swelling of the head), diarrhoea and respiratory distress. Birds will become seriously ill very quickly.

The symptoms of bird flu in a person are usually mild flu-like symptoms and mild conjunctivitis (sore eyes with a discharge). Bird flu virus H5N1 seems to differ from other bird flu types in its ability to cause serious illness with a high death rate in people. Bird flu can be treated with an antiviral medicine.

Professor Pennington, a microbiologist and President of the Society for General Microbiology has noted that the H5N1 virus has been around in China since 1996 and doesn't yet have the ability to pass from person to person. However, he also acknowledges that there is a definite theoretical possibility of the virus mutating, which could happen tomorrow, or not for another 20 years. Additionally, the professor highlights that we know far more about the virus now than we did in 1918 and we now have vaccines, antibiotics and antiviral drugs to combat its effects.

In the United Kingdom, DEFRA (Department for Environment Food and Rural Affairs) requires that anyone owning more than 50 birds must be registered with them, but anyone who keeps even just a couple of chickens is welcome to register with them. This would mean that you would be notified by text if anything occurred that might require you to take action. For information, visit www.defra.gov.uk

The Poultry Club of Great Britain has issued the following bio-security measures for anyone with free-range poultry:

- Wash hands before and after handling poultry
- Keep food under cover to minimise wild bird attraction
- Keep water fresh and free from droppings
- Keep waterfowl and chickens separate
- Control vermin
- Quarantine new stock for two to three weeks
- Change clothes and wash boots after visiting breeders
- Change clothes and wash boots before and after attending a sale
- Do not share chicken crates. If you must do so, disinfect before and after use
- Comply with import/export regulations
- Keep fresh disinfectant at the entrance to poultry areas for dipping footwear

Anyone with any suspicion that their birds have avian flu must contact their local veterinarian or the State Veterinary Service.

Laying difficulties

Do not panic if your hen stops laying – even the hybrid egg-laying machines do not lay daily throughout the year! You should make sure that you know the number of eggs your bird should lay approximately in a year. Your supplier should have given you this information when you purchased the bird. All hens will stop laying for one to two months, or longer, when they have their annual moult, see below. They automatically lay fewer eggs in the autumn and winter when days are short.

If a chicken's comb is not bright red then the bird will not be laying; this could also be an indicator of ill health, though it happens in the annual moult too. Look at her droppings, if they are formed, solid and dusted with white, then they are healthy, but if they are runny and a greenish brown she may be unwell. She may have eaten something that disagreed with her, but her comb will not be affected. Check the vent for worms and under the wings for lice and keep an eye on her over the next few days.

If you see any of the oviduct hanging out of the bird's vent she must be separated from the other birds and taken to the vet.

Moults

Birds lose their feathers annually in sequence beginning with the neck and the back, moving on to the breast and the petticoats and finally on to the back and tail. A young and perky bird will take around about six weeks over her moult, but an older hen can take up to three months. There is no specific time that this happens as it is dependent on the time of year that your hen hatched. Moulting is quite a dramatic process and your bird will look very sad and bedraggled. She will not lay and her comb will no longer be a vivid red. She will become very subdued and subservient to the other birds. They may well pick on her at this time, see Bullying on page 77. It is helpful to take your bird off her usual layers mix and put her onto food that contains more protein and carbohydrate to give her the extra boost she needs to grow her new feathers. Tonics are also available in liquid form to add to drinking water.

DEATH

The most common cause of death for a chicken is to be taken by a predator – in which case you will only know what has happened by some tell-tale clumps of feathers blowing across the garden and perhaps some drops of blood. If your chicken falls ill and dies suddenly, it is advisable to contact your vet in case there are any outbreaks of specific chicken diseases or viruses in the area that they need to monitor. Do not eat a chicken that has fallen ill and died.

PARASITES

Like all creatures, chickens have parasites that can affect them. If you think there may be a problem check your chicken. Look in the feathers around the neck, around the vent and under the wing. Also check the chicken house for any signs of infestation.

LICE

Like its infuriating relative the human head louse, the common fowl louse (*Menopon gallinae*) is a nuisance. Birds display clear symptoms – endlessly scratching themselves and pecking at themselves. The parasite can easily be detected by looking at the feathers close to the body; you will see clumps of white eggs and brown lice moving around, commonly, though not exclusively, in the crest and around the vent. Powders and sprays to treat the problem are available from vets and some pet shops and several treatments will be required.

MITES

These are one of the most common parasites to affect chickens. *Dermanyssus gallinae*, a red, eight-legged arachnid about 1mm long, lives in the cracks and crevices of the chicken house and emerges at night to feed on birds in the classic lousy fashion – they drink their blood. They can be seen in the edges and undersides of perches or other dark corners (simplest to spot at night with a torch); look for small clusters of red to blackish dots. Mites must not be ignored as they will affect your chicken's health and can carry other diseases. Clean and disinfect your chicken house thoroughly and leave removable parts exposed to sunlight for a few hours. When everything is clean and dry, dust all birds with a powder treatment just after they have retired for the evening and repeat after five to six weeks. Mites can live away from the birds for up to 36 weeks so you will need to be persistent to eradicate this pest.

SCALY LEGS

A chicken's leg should be smooth if it is lumpy and swollen she may have scaly leg mite (*Cnemidocoptes mutans*). A vet will be able to supply you with a preparation to help you treat this parasite, which lives under the scales in the leg, and advise how to deal with the scabs and crusts. If you suspect the problem has just developed, try dipping the bird's legs in surgical spirit every few days. These creatures thrive in damp conditions so the ventilation in your chicken house may not be sufficient.

WORMS

Chickens can pick up worm eggs very easily when rooting around the garden for food. If your bird seems off colour and her comb has faded, check her droppings and vent for any sign of worms. Vets sell worming treatments and some people automatically worm their birds twice a year.

PREDATORS AND PESTS

Chickens are not effective at seeing off predators – they have a tendency to panic! Some cats and dogs can be a bit of problem, though cats are generally OK with hens. Dogs should be on a lead when first introduced to your chickens to see how the dog reacts. Don't leave new birds alone with established pets until you are confident that there is no problem. It is not a good idea to have pet ferrets if you keep chickens, and mink and badgers can be a nuisance if they reside close by. Mice will be attracted by chicken food, clearing it away each night will discourage visitors, store food securely in lidded bins.

Fox

The fox (*Vulpes vulpes*) is the number one chicken predator. When a fox sees a chicken he will not give up easily in his quest to get at this delicious food. Chicken houses and runs must be fox proof if your chickens are to stand a chance. The fox is nocturnal, and he is an opportunistic scavenger eating birds, lambs, insects, worms, eggs, fruit and carrion.

Urban foxes are more brazen than their country cousins. The first indicator that they are present is often their terrible wailing screech – like a terrified child. This is often more frequent when they are mating in December to February. Cubs are born in March to May and foxes can be desperate for food around this time.

There is no easy solution to the threat posed by the fox other than having a secure chicken house and run. It is supposed to help deter the fox if human urine is sprayed around the borders of the property, which my husband and son take as an excuse to pee in the garden at will! Garden lighting does not discourage foxes at all.

If you want your chickens to be as free range as they can, then keeping your eyes and ears open to the threat and taking a risk or two is the only way you will find out the reality of the threat in your vicinity.

I have a very tough little chicken house and run and have witnessed the fox repeatedly attacking it in broad daylight in heavy snow. Mostly he leaves us alone in the day, but he visits the garden on a regular night time basis to check out the situation. He took one of my chickens, when I lived under the misapprehension that he would not visit in daylight. He killed poor Betty at dusk and took a chunk out of Wilma's wing (the primary flight feathers never grew back), leaving her traumatised and fearful for a good few weeks. He returned in broad daylight to take our very glamorous white chicken, Eileen, in broad daylight. Wilma escaped for a second time, obviously a wily old bird, and is still going strong.

RATS

Small gauge chicken wire and a secure chicken house are the simplest ways to deter rats. Ensure food is not left lying. If your chicken house is made of wood, check to see whether anything has been gnawing at it. Contact your local authority or a pest controller if you develop a real problem.

BIBLIOGRAPHY

COLLINS FIELD GUIDE
MAMMALS OF BRITAIN &
EUROPE by Priscilla Barrett and
David McDonald, *published by
Harper Collins*

PRACTICAL POULTRY KEEPING
by David Bland, *published by
The Crowood Press*

THE ENCYCLOPAEDIA OF
POULTRY by J.T. Brown,
*published by Walter Southwood
& Co Ltd.*

COLLINS FIELD GUIDE INSECTS
OF BRITAIN & NORTHERN
EUROPE by Michael Chinery,
published by Harper Collins

A GREENER LIFE by Clarissa
Dickson Wright and Johnny
Scott, *published by Kyle Cathie
Ltd*

ON THE FARM by Jimmy
Doherty, *published by Ebury
Press*

HOME POULTRY KEEPING by
Dr Geoffrey Eley, *published by
A&C Black Publishers Ltd*

THE EGG AND I by Betty
Macdonald, *published by
Hammond, Hammond & Co*

KEEPING PET CHICKENS by
Johannes Paul and William
Windham *published by Interpret
Publishing*

THE COMPLETE POULTRY BOOK
by W. Powell-Owen, *published
by Cassell and Company Ltd.*

KEEPING A FEW HENS IN YOUR
GARDEN by Francine
Raymond, *published by A
Kitchen Garden Book*

CHICKENS by Matthew Rice,
published by Matthew Rice

THE POULTRY MANUAL by Rev.
T.W. Sturges, *published by
Macdonald and Evans*

FREE RANGE POULTRY KEEPING
by Katie Thear, *published by
Framing Press Books*

THE COMPLETE
ENCYCLOPAEDIA OF CHICKENS
by Esther Verhoef and Aad
Rijs, *published by REBO*

THE PRACTICAL POULTRY
KEEPER by Lewis Wright,
*published by Cassell & Company
Ltd*

GLOSSARY

ALBUMEN The egg white

BANTAM A small fowl, distinct breeds, not a miniature version of a large breed

BARRING Stripes of light and dark colours running across a feather

BEAK The upper and lower mandibles of the chicken

BEARD A clump of feathers under the throat, appears in some specific breeds

BLOOD SPOT Blood within the egg yolk or white caused by rupture of blood vessels on ovulation

BOOTS Feathers coming out of the toes

BREED Fowl that breed true to a specific set of characteristics

BROILER Chickens of either sex under the age of nine weeks

BROODY The chicken equivalent of the ticking biological clock – they want to sit on the nest and hatch eggs

COCK A male bird over 12 months old

COCKEREL A male bird under 12 months old

COMB The colourful fleshy protuberance on the top of the head

CREST An upright tuft or crown of feathers on the top of the head

CROP The section of the gullet where food is held and prepared for digestion

FACE The skin all around the eyes

FLIGHT FEATHERS The wing feathers responsible for controlling flight, sub divided into primary and secondary sections

GIZZARD Food is contained here and partially digested

HEN A female bird over 12 months old

LACING A feather with a base colour in its body and a ring of a contrasting tone

LEG FEATHERS Feathers emerging from the outer side of the leg

LITTER Soft materials such as sawdust or straw used to cover the floor of the chicken house

MOULT Losing existing feathers and growing new ones

OVIDUCT The tube where ova are fertilised and eggs are formed

OVUM Unfertilised egg released from the ovary, plural ova

PENCILLING Small stripes or markings running across a feather, can also be described as bands

PEPPERING The down is a contrast colour to the main body of the feather

POINT OF LAY The term applied to young pullets of around 18 weeks old before they lay their first egg

PRIMARIES The stiff flight feathers at the outer tip of the wing

PULLET Female chicken under one year in age

ROOSTING Sleeping birds

SECONDARIES The large wing feathers closest to the body

SELF A feather in just one colour

SPANGLING A spot of colour on the end of the feather in contrast to its base colour

STRIPING A feather edged in a contrast colour

TIPPING A tiny v of contrast colour on the tip of the feather

UNDER COLOUR The colour of the downy plumage – principally the petticoats

VENT The exit point for both eggs and droppings

WATTLES The thin flaps that dangle from the base of the beak and throat

WING CLIPPING The habit of trimming the primary feathers on one wing to discourage the bird from flying

---

AUTHOR'S ACKNOWLEDGEMENTS

I would like to thank Sarah Gristwood and Fiona Lovering for their support and Emily Sutherland for her encouragement in the face of every fresh chicken flu scare. I am indebted to Polly Powell and Katie Cowan for their faith in me and to Miriam Hyslop for her gentle guidance. Mostly I would like to thank my family: Eric for his support, enthusiasm and the many chicken related cuttings, my daughter Florence for her superb proof reading, my son Teddy for his uplifting good humour, my stepdaughter Genevieve for her willingness to help a stressed author on the domestic front and my mother-in law, Lou, for telling me to crack on with my writing. Lastly I would like to thank the girls; Wilma, Dora, Betty, Lou and Eileen for their inspiration and eggs.

INDEX